**A TOOL FOR CONVICTED FELONS GETTING
STARTED IN A HVAC INDUSTRY**

HVAC
GUIDE

How to Remove a Residential HVAC System and
Install a New One

Joseph Johnson

Extreme Overflow Publishing
Dacula, Georgia
USA

For permission requests, contact the publisher.

Extreme Overflow Publishing

A Brand of Extreme Overflow Enterprises, Inc

P.O. Box 1811, Dacula, GA 30019

www.extremeoverflow.com

Send feedback to info@extremeoverflow.com

Printed in the United States of America

Library of Congress Catalog in-Publication

Data is available for this title.

ISBN: 979-8-9867000-9-0

Table of Contents

Table of Contents...2

Foreword 1

Introduction 4

Chapter 1 11

Tools and Materials 11

Chapter 2 18

The Process Step-By-Step..18

Chapter 3 35

Equipment In: Indoor Unit...35

Chapter 4 49

Equipment In: Outdoor Unit..49

Chapter 5 55

Releasing Refrigerant...55

Chapter 6 64

Clean Up...64

Conclusion..65

Bio 67

Foreword

The same belief I intuitively had about Joseph Johnson II five years ago, when I met him to record a podcast is the same belief I instinctively have about him today. He'll make it in this world at whatever he chooses to do. He has a way with people that creates results, a natural and good sense of humor that will go a long way, and a skillfulness to survive. These views are why his writing this book for you to learn about the HVAC trade is no true surprise. He has the life experience to write it for men who are starting from scratch (ask him his story) and the professional acumen as an HVAC technician to instruct anyone else who is entering the HVAC field.

This book provides a step-by-step, direct guide for beginners to learn about this lucrative business. It walks you through installation, disassembly, requirements, cleanup, and best practices. By taking a serious approach to this book, you can increase your professional knowledgeability and value, and change your life. In any event, that's not what I am here to go into detail about. Joseph will tell you everything you need to know ahead. I am here to tell you about what else you'll need to make it at this new venture. Drive, fight, spirit, and perspective. These four things will get you through a lot of it. The first three are similar to one another. If the four are in place, you'll make money and advance in this entrepreneurial way of life.

Drive is an innate urge to attain your goal. You need this because your drive is your in-house accountability partner, who keeps you dependable and adaptable. Fight, in this context, means to endeavor vigorously to win at something. If you want to see yourself as victor, you have to fight until you don't. You should reach a point where you win and can rest, but not before you go through the process to obtain the results. Keep going. Your spirit is the most important one of the four. According to Google's online dictionary, spirit is the nonphysical part of a person, which is the seat of emotions and character. How you feel and what you value is shown through your actions. Even if you think you are hiding where you stand, you are not. Thus, if you put something out into this world, it ought to be something that is good, something that contributes to your well-being and the well-being of the people around you.

Perspective is your point of view. Your ability to keep things in perspective will give you the levelheadedness you need to carry on. When judging your reality of what's happening, you'll need to be clear, honest, and prudent. Look at things from all angles to get a full picture.

Then thoughtfully proceed with your next action. If you can keep a grip around these assets, chances are, you'll gain staying power in your industry. Gut it out. You matter. Believe in yourself. Keep going, rest, and then keep going some more.

Ishna Hagan
Marketing Writer
Howard University Journalism Graduate

Introduction

According to the National Institute of Justice, almost 44% of released convicted felons return before their first year out of prison. The number one reason listed for people returning to a life of crime is difficulty in finding ADEQUATE employment. Think about that for a second. The main reason people get out of prison and go back to making the same decisions that got them put in prison in the first place is not being able to find a job that pays them enough to make better decisions. Convicted felons have to ask themselves two important questions about the job market if they want to give themselves a chance to get out and STAY OUT: 1) what skills do I have that someone will pay me for? and 2) will my criminal record stop me from getting hired? Committing to learning the HVAC trade will put you on the right side of both of those questions. I put this book together specifically for convicted felons looking for a sure-fire pathway to a successful career outside of illegal activity. Learning and mastering the information in this book will immediately make you an asset to your family and community. No matter where you go in the world, I assure you there's a need for heating, air conditioning, or both. There's just not enough people in the market to do the work. A convicted felon with the HVAC skills this book provides can make more money than a college graduate with a clean record within a year of being released. These days, trades pay much more than degrees. With that being said, let's start laying the foundation for your new life.

As a beginner in installing air conditioning systems, the most important piece of information you need is my Install Completion Checklist. You need this checklist in order for you to become efficient quickly. An Install Completion Checklist is the difference between your job as an HVAC Install technician and a lucrative career as an HVAC Install Technician. With this list, you'll be able to review the tasks that must be checked off before an a/c install can be considered complete which puts more money in your pocket.If someone were to approach you right now and say, "Hey, I'll pay you $3,500, plus the price of materials for you to put a new air conditioning system in for me. Is that alright?" The Install Completion Checklist will help you determine exactly what you need to have in place to put that $3,500 in your pocket! The great thing about this list is that even if you work for a company, the Install Completion Checklist will help you understand exactly what you're being paid to do.

As a beginner in the HVAC field, you should be looking to answer three questions from day one:

(1) Exactly what tasks am I going to be expected to complete?

(2) What tools do I need to complete these tasks?

(3) How exactly do I complete these tasks?

The objective of this book is to answer these three questions for you so that you can begin to master a recession-proof skillset.

First, I'll provide you with a universal Install Completion Checklist, which will cover all of the tasks you'll be expected to complete as a beginner. The more efficient you are at these tasks, the likelier you'll be viewed as more than a beginner. Next, I'll provide you with a fully detailed tool list. Your tools dictate how efficient you are at completing the necessary tasks and have a direct effect on how much employers will pay you. You can literally negotiate how much a company pays you, even if you're not that good, as long as you have all of the necessary tools.

I'll tell you exactly what tools to get and the best place to get them from, so you are viewed as a bit more than an amateur, even though you actually are a beginner. Thirdly, I'll go through each of the required tasks step-by-step and explain exactly how to get it done and what tools you'll need to get them done. Without further ado, let's start with the Install Completion Checklist.

You don't have to follow it in the exact order I wrote it, but you do have to make sure each of these tasks are completed before you say the job is done.

- Recover refrigerant or pump down the unit
- Disassemble and remove the AHU
- Disassemble and remove the CU
- Set and level the new AHU
- Set and level the new CU and its pad
- Reattach the ductwork to the new AHU
- Complete low voltage wiring at the AHU

- Complete high voltage wiring at the AHU

- Run the drain line

- Fit the refrigerant lines back into the AHU

- Fit the refrigerant lines back into the CU

- Complete low voltage wiring at the CU

- Complete high voltage wiring at the CU

- Seal the wall penetration with silicone or foam

- Braze refrigerant lines

- Vacuum the refrigerant lines

- Release refrigerant

- Check refrigerant numbers and charge unit if necessary

- Clean up

Keep in mind that installing an a/c system is a two-man job, so these tasks will typically be divided between you and one other person, very likely more skilled than yourself. Certain tasks like brazing, checking refrigerant numbers, and charging the unit with refrigerant are all tasks that'll typically be performed by someone a little more knowledgeable than a beginner.

Your role as a beginner is to complete the tasks that a beginner should be able to handle with no problems, in a timely manner, and with little to no supervision.

Now is a good time to point out a small but important difference in installing air conditioning systems in different parts of the country. I've installed air conditioning systems in the humid, warm-weathered state of Florida, as well as the dry cold-weathered state of Colorado. For the sake of simplicity, in this book I'll be explaining the process of installing an air conditioning system the way it would be done in a warmer state, like Florida. The difference between installing air conditioning in a warm weather state and a cold weather state is what the indoor unit is physically attached to. Your understanding of this point will become clearer in Chapter 5 when we go over the process of re-attaching the new indoor unit to the existing ductwork.

In warm-weathered states like Florida, there is not a big emphasis on heating, so they typically install air handler units there. Air handler units are made up of three major components: the evaporator coil, which is the bottom portion of the AHU; the blower fan, which is the middle portion of the AHU; and the heat strip, which is usually inside the wiring panel at the top of the AHU. In cold-weather states like Colorado, there's a much bigger need for heating so they use gas furnaces. When gas furnaces are present, you'll only need to attach an evaporator coil to it rather than installing a full AHU. All this means is if there's a gas furnace present, then you don't have to worry about wiring any high voltage at the indoor unit. Other than that, the process is the exact same. One of the first things you need to figure out is whether the homes in the city or state where you intend to work typically have gas furnaces or air handler units. If the system has a gas line going to it, it's a furnace. If you can literally see a flame inside the system when you turn the heat on, then it's a furnace. If you don't see a literal flame when you turn on the heat, then it's most likely an AHU. Now, before we jump into the actual process of installing the equipment, let's get familiar with the lingo of the HVAC industry.

Chapter 1

Tools and Materials

Materials

Residential and commercial HVAC for the most part share the same materials, but there are some differences.

The primary materials for the indoor unit are the air handler unit (AHU) itself, which consists of the evaporator coil, the blower fan, the heat strip, the filter, the ductwork, and the thermostat.

The primary materials for the outdoor unit are the condenser unit (CU) itself, the pad, and the disconnect box. Lastly, there's what I call linking materials which include the thermostat wire, the condenser wire, and the refrigerant line set. I refer to these three materials as linking materials because they link one piece of equipment to another piece of equipment.

Tools

This is a list of tools you'll need for every job (or at least most jobs):

- 2 medium-sized tool bags – H.D.

- 1 bookbag – H.D.

- Duct knife – H.D.

- Impact gun/drill – H.D.

- Copper cutters – H.D.

- Wire strippers – H.D.

- PVC cutters – H.D.

- 5/16" chuck – H.D.

- Red & green offset tin snips – HVAC

- Linemen or Dikes – H.D.

- Service wrench – HVAC

- Set of Allen wrenches – H.D.

- 11-in-1 multi driver – H.D.

- A pair of crescent wrenches – H.D.

- Thermostat screwdriver – HVAC

- Attachable Phillips bit for drill – H.D.

- Attachable 90-degree bit for drill – H.D.

- Multiple length extensions for drill – H.D.

- Torpedo level – H.D.

- Copper benders – HVAC

- Multimeter – H.D.

- ¼ chuck – H.D.

- Flathead screwdriver – H.D.

- Service wrench hex adapter – HVAC

- Phillips screwdriver – H.D.

- A pair of channel locks – H.D.

- Hammer – H.D.

- Pry bar – H.D.

- Tape measurer – H.D.

- Sharpie markers – H.D.

- Small tool pouch – H.D.

It's really important to have quality tools. With that in mind, if the tool has HVAC beside it, then you should only get that tool from an HVAC store because that's the only place you'll get the quality you need for that particular tool. If it has an H.D. beside it, then it's okay to get that tool from Home Depot or Lowe's. Obviously, though, all of these tools are available at both HVAC and department stores.

Other tools that are important but not used for every single job are:

- Seamers – HVAC

- A pair of vise grips – H.D.

- 12" or 24" folding tool – HVAC

- Crimpers – HVAC

- Flashlight/headlight – H.D.

This is a short description of each tool:

- 2 medium size tool bags = self-explanatory

- 1 book bag = self-explanatory

- Duct knife = regular looking knife but specifically made for cutting ductwork

- Phillips screwdriver = self-explanatory

- Flathead screwdriver = self-explanatory

- Copper cutters = small little hand tool that has a really small circular blade with two little rolling wheels that's used for cutting refrigerant lines

- Wire strippers = hand tool used for cutting or stripping wires

- Impact drill = screw gun

- chuck = attaches to the impact drill and used to remove screws

- ¼ chuck = attaches to the impact drill and used to remove ¼ screws

- Red & green tin snips = medium sized hand tool used for cutting sheet metal. Red handled tin snips are for cutting with your right hand. Green handled tin snips are for cutting with your left hand. There will be times when you have to cut in both directions. Both red and green tin snips MUST ABSOLUTELY be offsets. DO NOT GET STRAIGHT CUT SNIPS!

- Linemen/Dikes = medium sized hand tool that looks like larger grip pliers, but they can also cut wires

- PVC cutters = used for cutting PVC pipe

- Service wrench = small hand tool. Difficult to describe with words but used for opening and

- closing king valves

- Service wrench hex adapter = attaches to the service wrench to assist with opening and closing the king valves

- 11-in-1 multi-driver = screwdriver with 11 different changeable heads

- 2 pair of channel locks = self-explanatory

- A pair of crescent wrenches = self-explanatory

- Thermostat screwdriver = really small screwdriver used for wiring the thermostat

- Attachable Phillips bit for drill = self-explanatory

- Attachable 90-degree bit for drill = this drill bit attaches to your impact drill and allows your drill to access screws in tight places

- Multiple length extensions for drill = these drill attachments allow your drill to access screws that are a lot further away

- Pry bar = self-explanatory

- Torpedo level = small 6" hand tool used to make sure equipment is level

- Hammer = self-explanatory

- Tape measure = self-explanatory

- Sharpies/markers = self-explanatory

- Small tool pouch = self-explanatory

- Set of Allen wrenches = self-explanatory

- Copper benders = large hand tool used for bending refrigerant lines. You'll have to be taught how to use this tool.

- Seamers = medium-sized hand tool used to fold or bend sheet metal

- Crimpers = medium-sized hand tool specifically used to crimp round duct

- Vise grips = self-explanatory

- Flashlight/headlight = self-explanatory

- Multimeter = medium sized hand tool used to measure voltage, amperage, and continuity

Chapter 2

The Process Step-By-Step

The steps laid out here describe exactly what you need to do to complete what's typically referred to in the industry as a retrofit. A retrofit is the process of removing a unit that is no longer functional and installing a new one. Being able to execute these steps with very little to no supervision will immediately prove you deserve to be paid a much higher wage than a novice. The more you know and the more tasks you can do without supervision, the more you get paid. It's really as simple as that. Now, let's start working on getting you paid!

First, the old Air Handler Unit and Condenser Unit have to be disassembled and removed.

Second, the new AHU and CU have to be set into place, reattached, and leveled.

Thirdly, the refrigerant lines need to be dry fitted into the units and brazed in. Next, the system has to be pressure tested. While that is happening, all of the wiring should be taking place.

The thermostat should be mounted if a new one is required. From there, it's time to vacuum the system. After the vacuum holds below 500 microns, it's time to release refrigerant and do final tune ups.

Disassembly

Disassembly is the total process of completely removing the AHU and the CU. Once the AHU and CU are 100 percent disassembled, you should be able to load them onto a truck and haul them away. They should be 100 percent detached from any equipment that is going to be reattached to the new AHU and CU and/or anything connected to the home.

The first objective of disassembly is to kill the high voltage power to the AHU and the CU. To accomplish this, let's start with the AHU (or indoor unit). You need to locate what's called the main breaker panel. Main breaker panels are rarely in the same location from home to home. To save time from searching for it, it is standard to just ask the homeowner where the main breaker panel is located. Once you locate the main breaker panel, you'll need to identify the breaker labeled a/c. Flip it to off. It is not a universal law that all breaker panels are labeled correctly, which is why you'll need a multimeter. Sometimes the breaker for the AHU is labeled heat or can even be mislabeled altogether, so the next step is to confirm that you've killed power to the AHU. This is accomplished by opening up the wiring panel on the AHU and identifying the breaker inside.

The breaker inside the wiring panel of the AHU is what brings high-voltage power from the main breaker panel to the AHU itself, then distributes that high-voltage power to different components within the AHU. You need to use your multimeter to verify there's no power going INTO the breaker inside the AHU. You do this by putting your multimeter on the voltage setting, then placing one lead of your multimeter on one of the breaker terminals and your other lead of your multimeter on any piece of metal inside the AHU. Once your multimeter reads 0.0v coming from the breaker inside the AHU, you're good to go. If you are still reading voltage at the breaker inside the AHU on your multimeter, you'll have to go back to the main breaker panel and flip each breaker off one by one until your multimeter reads 0.0v at the breaker inside the AHU. No power should be coming into or going out of the breaker inside the AHU.

Once you've confirmed there's no power at the breaker inside the AHU, the next step is to disconnect the line-side wiring, which are the three wires coming from the main breaker panel to the breaker inside the AHU. You'll need to use either your flathead screwdriver, your Phillips head screwdriver, or your 11-in-1 multi-driver to release the wires from the screws. The three wires you're detaching are the hot leg, the neutral leg, and the ground wire of the line-side voltage, which is the technical name for the wires coming from the main breaker panel to the breaker inside the AHU.

Now go to the outdoor unit. The very first decision that has to be made when you get to the outdoor unit is how are you going to get the actual refrigerant out of the refrigerant lines? There are only two possible answers to this question, and the answer is solely determined by what tools you have. The process of removing refrigerant from the system is known as evacuating the system. The tools you need to evacuate a system are a recovery machine, a set of digital or manifold gauges, a recovery tank, and a few hand tools. As a beginner, you most likely won't be handling these tools until someone teaches you how to use them hands on. Because you don't know how to operate the tools to recover the refrigerant in this way, you have only one option—to perform what's called a pump down.

Pumping down the unit is another acceptable way to evacuate the system. The objective of pumping down the unit is to trap all of the refrigerant inside the compressor. The compressor is a component of the condenser unit that is capable of holding and storing all of the refrigerant in the refrigerant lines. To accomplish this, you need a set of digital gauges, a service wrench, a service wrench hex key adapter, and a screwdriver. It's important that the outdoor unit still has power. You'll also need to know what king valves are, how to open and close them, and how to remove the king valve caps.

Use your channel locks or a crescent wrench to remove both king valve caps. Second, hook your gauges up to the liquid line and the suction line at the CU. You need the gauges because they'll tell you when all of the refrigerant is out of the refrigerant lines and trapped inside of the compressor. Once your gauges are hooked up to the system, the next step is to close the king valve on the liquid line. Doing this ensures that no refrigerant will leave the compressor. To close the king valve on the liquid line, you'll need to use your service wrench or the appropriately sized Allen wrench. Stick the service wrench or Allen wrench in the top of the king valve and turn it clockwise until it won't turn anymore. When the king valve won't turn anymore, that means the liquid line is completely closed. You should be able to look into the top of the king valve and see that you've wound it down. Now the next step is where the work is done. The objective of this next step is to use the compressor to trap the refrigerant. This is a good time to give a brief overview of the refrigerant cycle. The refrigerant starts off inside the coils of the condenser unit. It travels from the condenser unit coils through the liquid line all the way to the metering device which is located inside the AHU. The refrigerant passes through the metering device and into the evaporator coil. From the evaporator coil, it travels through the suction line to the compressor and the cycle continues repeatedly. But now that we have the liquid line closed off, the refrigerant won't be able to cycle through the system.

To trap the refrigerant, you'll need to engage the contactor by pushing it in with a screwdriver. This step can be somewhat dangerous, so be careful when pushing the contactor in because the contactor is passing high voltage through it. As soon as you push the contactor in, you should immediately hear the CU kick on. You should already have your service wrench (with the service wrench hex key adapter) or the appropriately sized Allen wrench positioned inside the top of the king valve of the suction line and anticipating the right time to close it off, exactly like you did the king valve on the liquid line. While you are pushing the contactor in, you need to be watching the refrigerant pressure numbers on your gauges. The refrigerant pressure numbers will be consistently falling towards zero. Basically, what's happening is the compressor is sucking the refrigerant into itself, then letting some of the refrigerant flow into the condenser coil. The refrigerant has nowhere else to go from there because the liquid line king valve (which serves as a gateway) is closed. Keep the contactor engaged until your gauges read zero on the refrigerant pressure for both the liquid and suction lines. As soon as your gauges begin to read five psi on your suction line, you should start closing the king valve on the suction line. Once your gauges read zero on the suction line pressure, you should close that king valve completely.

Now the vast majority of the refrigerant is trapped inside the compressor and a little bit is inside the condenser coil. At this point, you've successfully completed a pump down. Great job! Now it's time to get that condenser unit totally detached. The very next thing you need to do at this point is shut off the power at the outdoor disconnect.

Outdoor disconnect boxes come in all shapes and sizes, but they all serve the same function. They just carry high voltage power to the condenser unit in the same way the main breaker panel carries high voltage power to the AHU. Ninety-nine percent of the time the disconnect box for the outdoor unit will be located on the outside wall of the house, no more than ten feet from the unit. Most times, it's literally right next to the unit. Disconnect boxes come in several different styles, so to shut off the power, you may have to pull the handle down or flip a switch or even pull the plastic connector piece out. Either way, shutting off a disconnect is something you can figure out on your own with no problem. To double check to make sure you killed power to the condenser unit, press the contactor in again. If nothing happens, then you've successfully killed power. If the CU jumps on when you engage the contactor, then you have NOT successfully turned off power at the disconnect box. At this point, you should just save time and ask the homeowner where is the disconnect or breaker for the outdoor unit. Once you've confirmed there's no power going to the unit, you're ready to rock and roll.

At this point, you want to cut the refrigerant lines. Use your copper cutters to cut the lines a few inches before the king valves (say five to eight inches). Then use your Phillips screwdriver to back the screws out of the part of the contactor that's holding your line-side voltage. You should be able to physically see and follow the three wires coming from the disconnect box and going to the condenser unit. You just have to make sure you detach those wires from the CU. After that, you have to detach your two low-voltage wires. Depending on the brand of equipment you're working with, these two wires could be located in either of two spots. Lennox, Carrier, Daikin, and Goodman brands all have their low-voltage wiring set up a little differently. The easiest way to go about this is to physically trace the wire with your eyes and hands. If you look where the refrigerant lines penetrate the house, you should see a small, brown-sheathed wire running alongside the refrigerant lines and up into the condenser unit. Inside the unit, you'll realize the brown-sheathed wire is actually the protective outer casing for two small red and white wires. These two wires are your low-voltage wires for the CU. Often they're referred to as condenser wire or just two-wire. Use your wire strippers to simply cut the low-voltage wires anywhere inside the CU. At this point, the CU is ready to be scooped up with a hand dolly and carried away. Disassembly of the CU is now complete.

Now let's go back inside to the AHU. You've already disconnected the high-voltage power for the AHU, so right now your thought process should be solely locked in on detaching the AHU from the components of the system that are staying put. This is relatively easy to accomplish once you identify what components are staying and what components are leaving, which we'll spell out later in this chapter. For now, let's just focus on the process.

The first step you can get done quickly is cutting the refrigerant lines. The same way you cut the refrigerant lines at the CU, you're going to cut them at the AHU. To be safe, cut them five to eight inches from where they enter the AHU. At this point, the refrigerant lines should not be connected to anything inside or outside.

Now focus on the low-voltage wiring inside the AHU. There are two sets of low-voltage wiring that you're looking to disconnect. You should be able to easily locate both sets of low- voltage wires inside the AHU at what's known as the control board. Remember the brown-sheathed low voltage wire you cut with your wire strippers at the CU? The one I referred to as two-wire or condenser wire that runs alongside the refrigerant lines? Well, the other end of that wire is what you're going to disconnect from the control board inside the AHU. Physically follow that wire with your eyes and fingers from anywhere outside the AHU to inside the AHU, and it should lead you directly to the two-wire anywhere inside the AHU. Keep in mind you're going to have to re-use that same two-wire later, so try to leave some extra slack if you can.

Next, on that same control board, you should see five terminals labeled R, C, Y, W, and G with a set of wires going to each terminal. This set of wires is how the thermostat communicates with the AHU. This set of wires is also known as thermostat wire. Depending on the type of unit you're working with, this thermostat wire could have up to eight wires. Some of these wires may very likely be wire-nutted to other wires, but that's irrelevant because all we're concerned about is saving enough of the thermostat wire to reconnect to the new AHU. You can literally cut this thermostat wire anywhere inside the AHU, but you should be thinking two things before you cut the wire:

(1) when I cut this wire, will this thermostat wire be completely detached from any and every component inside the AHU? And

(2) will you still have enough slack in the wire, so it'll be able to reach the new control board inside the new AHU?

The next step is the condensation drain line.

The evaporator coil inside the AHU has to have a way to drain excess moisture/water. It does that through a piece of " PVC plastic pipe that goes directly from the AHU to a floor drain. Typically, the floor drain is in the same room or space as the AHU, but that's not the case every single time. You're going to use your PVC cutters to literally cut the " PVC pipe that's being used as a condensation drain line. At this point, you know what to cut (" PVC drain line) and how to cut it (PVC cutters). Now you should be deciding where to cut it. If the floor drain (which is where the PVC pipe should terminate) is in the same room or space as the AHU, then it doesn't matter where you cut it because it's all going away. If the entire piece of PVC pipe (the entire drain line) from the beginning of it, which is at the AHU, to the end of it, which is at the floor drain, is all inside the same closet space, or room, then you're going to get rid of all of it. If the drain line actually goes through a wall, ceiling, or floor, then you'll only be getting rid of most of it. To say it in a different way, if you can't physically see where the drain line terminates from the room or space the AHU is sitting in, then you're going to have to leave a small portion of the drain line to reattach to. You should be asking yourself this question: where exactly can I cut this drain line so that I get rid of the part that's attached to the AHU and most of the rest of it but still leave enough so that I can reattach a new drain line when it's time to? A general rule of thumb is to cut the drain line six inches away from where the drain line is penetrating the wall, floor, or ceiling. This is something you just need to use your own discretion and common sense with.

The only task remaining to complete the disassembly step is to physically detach the AHU from the ductwork itself.

The ductwork will be made of either sheet metal or fiberglass foam, and it can be attached with either screws, tape, mastic, or any combination of the three. This is not always the case, but most times there's ductwork attached to both ends of the AHU, the top and the bottom. Your objective is to detach the AHU from the ductwork completely without destroying the ductwork because the new AHU is going to have to be reattached to that same existing ductwork. If the ductwork is made of fiberglass foam, then it should only be attached to the AHU with tape and/or mastic. In this instance, you just need your duct knife. You'll want to save as much ductwork as possible, so position your duct knife around the top of the tape on the ductwork, which should only be a few inches above the top of the AHU. You'll need to cut as straight a line as you can horizontally, all the way around the ductwork, until the ductwork is completely detached from the AHU. Fiberglass foam is 1" thick typically, so you definitely have to use a little force when cutting it.

If the duct work is made of sheet metal, then it should be attached to the AHU with screws. However, sometimes it may be attached with tape or mastic or even all three. Use your impact drill to remove the screws. Typically, the screws are either 1/4" or 5/16", which is why you must have a 1/4" and a 5/16" drill bit (aka Chuck). After you remove the screws that attach the metal ductwork to the AHU, or peel the tape off, or pry the mastic loose, the AHU and the metal ductwork should become detached. Sometimes it is necessary to actually cut the metal ductwork away from the AHU. This is where your red and green offset tin snips come into play. Where you should begin cutting is really on a case-by-case basis, so you have to use your own discretion while keeping in mind that you want to save as much ductwork as possible. Again, you want to cut as straight a line as possible horizontally, all the way around the ductwork, just like you would do if the ductwork were fiberglass foam. The only difference is obviously, you can't use a duct knife to cut sheet metal like you would on fiberglass duct. You can use your tape measurer and Sharpie to measure out and mark a straight line on the sheet metal where you want to cut along. However, before you begin cutting with your tin snips, you'll have to make an incision in the sheet metal that your snips can actually fit into. You could use a step-bit that attaches to your drill to make a hole big enough for your snips to fit in. If you don't have a step-bit, you can use your flathead screwdriver on the sheet metal sideways so that only one corner of the flathead tip is touching the sheet metal. Bang the tip of the screwdriver with the hammer. After a few good knocks, the tip of the screwdriver will pierce the sheet metal, creating just enough space to operate your tin snips. Once you cut the sheet metal all the way around, the AHU should be completely detached and ready to be loaded onto a truck and hauled away. There should be absolutely nothing attached to the AHU unless it's going away.

The last point I want to make about disassembly is every job you do may have slight variations, but the process is always the same. The process is always to get the refrigerant out, then detach all of the equipment that's going away from all of the equipment that is staying. In a warm state like Florida, all you have is the AHU. You can follow this disassembly outline step- by-step to reach your desired results. In a state like Colorado that has a much colder climate at times, they don't have AHUs, they have gas furnaces that have air conditioning installed on them. Disassembly of an air conditioning system attached to a gas furnace is the same exact process minus the wiring. When an air conditioning system is attached to a gas furnace, you follow the exact same steps. The only difference is, at the INDOOR UNIT, you don't need to disassemble any wiring.

Chapter 3

Equipment In: Indoor Unit

N ow that the old AHU is completely removed, the next step is to set, level, and attach the new indoor unit. The very first question you have to ask yourself is, what exactly is the new unit going to sit on? All scenarios are not the same in this regard, though. Some AHUs may sit on top of a wooden platform with a square hole cut out of it inside of a closet. Or, some units may sit on top of a metal rack. And even still, some units may require you to build what's called a plenum box for the unit to sit on. Building plenum boxes isn't something you'd be expected to be able to do as a beginner, so I'm going to proceed as though you're just setting the new AHU on top of a wooden platform.

Now that you've determined what the AHU is going to sit on, the next thing to figure out is if the platform you're going to sit the AHU on is leveled. To determine that, you just need to place your torpedo level flat on top of the platform and read the level. This will tell you whether or not the AHU is going to be level, and if not, what side needs to be raised.

Once that is settled, you'll need to grab your tape measurer and measure the distance between the top of the platform and the bottom of the existing ductwork. Remember in the "DISASSEMBLY" chapter when you cut the ductwork away from the old AHU? Well, now you're about to put the new AHU in the same exact spot, but the old one and the new one may not necessarily be the same height, so now is the time where you determine exactly how much space you need for the new AHU. Measure the new AHU from the very bottom of the unit all the way up to the lip/indention of the unit. This will obviously tell you how tall the new unit is. Once you know how tall the new unit is, now you know exactly how much space you need between the top of the platform and the bottom of the ductwork. So, let's just say you measured the height of the new unit to be thirty inches. If the bottom of the ductwork is only twenty-seven inches from the platform, then you know you have to trim three more inches off of the ductwork, so it'll fit snuggly onto the AHU.

After you get the space between the platform and the ductwork to fit the new AHU perfectly, it's time to set the unit in place. You'll most likely need someone to help you pick up the unit to get it in place. Once you get the new AHU set exactly where you want it, it's time to begin reattaching everything.

First, start with the ductwork. There's a lot more than one way to skin a cat when it comes to reattaching ductwork, and as a beginner, this is something you should learn hands on. The primary objective is to secure the existing ductwork to the new AHU. If the steps to accomplish that seem intuitive, then go for it, but if it doesn't seem obvious as to how to secure the ductwork to the AHU, then you should leave that task for someone with a little more experience.

While someone else is working on reattaching the ductwork, the next task you can complete is wiring the high voltage back in. Remember, during the disassembly process, when you detached the hot leg, the neutral leg, and the ground wire of the line-side voltage? It's now time to reattach those wires. If the person that's responsible for providing all of the correct materials to complete the a/c install did their job correctly, then you should have a correctly sized breaker on hand that is ready to be attached inside the wiring panel of the AHU. Figuring out exactly how and where to put the breaker is something a beginner can figure out, so use your intuition there. Once you secure the breaker into place, you're going to attach two of the three high-voltage wires to the breaker. My way is not the only way to do this, but this is exactly how I wire up the voltage. I use my Phillips head screwdriver to open up the screw on the top, left-hand corner in the front of the breaker. I typically open it enough to fit the exposed wire of the hot leg into it, then tighten it up. I do the same thing with the screw on the right of the breaker. The only difference is I slide the neutral leg into that one. Lastly, you have the ground wire. It's super important to always get the ground wire right, which is extremely easy to do. The ground wire will always be the green one or the only one that's fully exposed. In the trio of high voltage wires, the hot and neutral legs are typically black and red, or black and white, or sometimes even both black. Regardless of what color wires are present, the green wire will always be the ground wire. If there's no green wire present, then the ground wire will be the only exposed, bare wire there. Green wire or bare wire are the only two ways I've seen ground wire. The ground wire typically has its own terminal off to the side. You need a flathead screwdriver to open and close the screw for the ground terminal, unlike the Phillips, which was necessary for the hot and neutral terminals. Once you get those three wires in, your high-voltage wiring should be complete. If there's additional high-voltage wiring that needs to take place, then you should step aside, let someone more experienced finish the high-voltage wiring and move on to the next task.

At this point, you should be making a mental note of exactly what work is remaining for the indoor unit. A list of your remaining work looks like this: (a) fit refrigerant lines into the unit, (b) build the condensation drain, (c) complete low-voltage wiring, (d) mastic and, if necessary, the thermostat replacement. There's no particular order these tasks HAVE to be completed in. It's totally up to you as to what order you want to tackle these tasks, but I'll address them in the order I do them.

With that being said, it's time to move on to fitting the refrigerant lines back into the new AHU. This is where your copper benders come into play. Right now, you need to determine what size the existing refrigerant lines are and what size are the refrigerant lines on the AHU. The objective is to get the existing refrigerant lines to fit back into the AHU. Let's start with the suction line, which will ALWAYS be the bigger of the two copper refrigerant lines and will be the one that is insulated. Like I mentioned previously, the first thing you need to figure out is the size of the existing suction line (which is always the bigger, insulated refrigerant line) and the size of the suction line on the evaporator coil (which is obviously in the AHU). Suction line sizes can range from 5/8" to 1" and maybe even bigger. It literally just depends on the equipment and the brand. I've been working with Lennox and Daikin equipment the most over the past year or so, so I'm going to use their residential equipment as a reference point. Typically, suction line sizes are 3/4" or 7/8". There are multiple ways to determine the size of the suction line, so I'll list a few. One way is to look at the insulation around the suction line. The insulation will always have its size printed on it, and obviously the size of the insulation correlates with the size of the suction line it's on. The only problem with this method is that over time, the print on the insulation becomes faded, so it may or may not be readable. If it isn't readable, then you can always use this next method. Simply get a copper fitting, either a coupler or a ninety-degree elbow and fit it onto the suction line and see if it fits. Literally take a 7/8" copper fitting and place it on the end of the existing suction line. If it fits onto it snuggly, then that entire suction line is 7/8" in size. If it doesn't fit snuggly, then try a different size fitting. The most common suction line sizes I see in the field are 3/4" and 7/8" so those are always the first two sizes I try. This is a surefire way to determine the size of the suction line. You or whoever you're working with should always have multiple sized copper fittings on hand. Another surefire way to determine the size of the suction line is by using your copper benders. Your copper benders will come with a piece called a bending mandrel. Actually, it will come with several different sized bending mandrels from 1/4" – 7/8". You can literally place different sized bending mandrels on the existing suction line until one fits on nice and snug. Again, I'd start with the 3/4" or 7/8" mandrels.

Once you determine the size of the existing suction line and the suction line of the evaporator coil, it's time to complete that connection. Right now, you should be thinking, what do I need to do to make the existing suction line fit into the evaporator coil suction line? Most likely, you're going to have to add copper fittings and a piece of copper pipe to the existing suction line to accomplish this. Use your tape measurer to measure the distance between the end of the existing suction line and the start of the evaporator coil suction line. This will tell you exactly how much copper pipe you'll need to get from point A to point B. It's totally up to you how you make that connection happen, too. You may need to use a combination of ninety degree elbows, couplers, and reducers, or you may need to use your benders to get the existing copper line at the correct angle to fit perfectly into the evaporator coil suction line. This is a task where you just have to look at exactly what needs to be done, then figure out how you can get it done. Keep in mind while you're fitting the copper together that you need to do it in such a way that the front panel of the evaporator coil can still be taken off and put back on at any time.

Once you get the suction line fit in, the next step is fitting in the liquid line. This is always the easier one because the liquid line is always a lot smaller than the suction line, and it can be bent by hand. The liquid line is usually 1/4" or 3/8". Again, you can determine the size by using a copper fitting. Place a 3/8" coupler on the end of the existing liquid line and see if it fits. When you determine the size of the liquid line, use your tape measurer to measure the distance between the end of the existing liquid line and the start of the liquid line on the evaporator coil. From there, you should be able to put a coupler on the end of the existing liquid line, cut a piece of copper pipe long enough to connect the existing liquid line to the liquid line on the evaporator coil using your copper cutters, and bend the line into place by hand. At this point, you should have both refrigerant lines fit into the evaporator coil.

The next task to complete is the condensation drain line. Nine times out of ten, the condensation drain line is run in 3/4" PVC pipe. The evaporator coil will always have two parts connected to the drain pan, specifically for 3/4" PVC to attach to. These two ports typically have a red plug in the left port and a green plug in the right port. Depending on the brand of the AHU, it may or may not have a green plug in the right port, but all brands have a red plug in the left port. Some brands position these ports in different locations sometimes, so the important thing you need to remember is NEVER run your drain line on the port with the red plug in it. We'll address what that port is used for momentarily. For now, you just need to identify the condensation drain port that has a green plug in it or no plug at all. That's the part we're going to begin with. The very next thing you need to do is locate the floor drain that should be inside the same room as the AHU. Remember, during the disassembly phase, we identified where the condensation drain termination was. It was either at a floor drain in the same room as the AHU or in another room somewhere, in which instance you'd have cut the existing condensation drain line about six inches from where it penetrates the wall, floor, or ceiling. It may likely be tied into what's known as a condensation pump. Your objective at this point is to run the 3/4" PVC from the drain port at the evaporator coil to the floor drain inside the room, to the six-inch piece of PVC remaining from the existing drain line or to the condensation pump. You want the drain line to look uniform and tight to the AHU. You'll most likely have to run some portions of the drain line along the floor, so keep in mind you need to run it in a way that it doesn't become a tripping hazard for the homeowners. Start with a 3/4" male thread fitting. It's the fitting that's threaded on one side and allows you to slide a piece of 3/4" PVC into it. Attach the " male thread fitting to the drain port on the AHU. From there, you have to use your discretion to figure out the best way to get the PVC from the male thread fitting to the floor drain, condensate pump, or the piece of existing PVC pipe. There are two things you need to keep in mind when running the condensation line. Number one, you must run it so that it's pitched away from the male thread fitting so that the water will flow downstream. You can't have the water getting backed up into the AHU, so you have to make sure the PVC is angled in a way that the water can flow downwards. Secondly, you have to add what's called a vent to the drain line.

To do this, you have to put a tee anywhere on the drain line facing upward so that the drain line can breathe, so to speak. If you've ever seen or heard of someone pouring bleach into their drain line, they were pouring it into the drain line vent. Keep in mind that water will be flowing through this drain line. You have to position the PVC tee so that water can't spill out of it because you're going to leave the top of the tee open. You'll use blue lava PVC glue to glue the PVC together.

Now let's look at the drain port with the red plug. This is where the SS2 float switch goes. The SS2 float switch is a safety device that turns the a/c off when it detects too much water. If the condensation drain line gets backed up, it'll begin funneling water into the SS2 float switch. Once enough water gets in there, it'll cut power to the a/c so that the a/c will stop producing water and prevent flooding. The SS2 float switch comes with three components: a male thread fitting, a ninety-degree elbow, and a cap with a rubber buoy and two black wires attached to it. So first, attach the male thread fitting to the drain port right next to the condensation drain line, the one with the red plug in it. Next, attach the ninety-degree elbow to the male thread fitting. From there, put the cap with the buoy and two wires onto the top of the ninety-degree elbow. Now those two black wires have to be run to the control board inside the wiring panel of the AHU.

This is the time to complete the low-voltage wiring for the AHU. There are three sets of low-voltage wires that need to be wired up at the AHU and they all get wired into the control board inside the wire panel of the AHU. This means you should have the five wires from the thermostat, the two wires coming from the condenser unit running right along the refrigerant lines, and the two black wires from the SS2 float switch. Start with the five wires from the thermostat. The objective is to make sure that the wires inside the thermostat correlate with the thermostat wires on the control board. The smartest way to do this is to actually go to the thermostat and pull the face off, which will reveal the wiring terminal. You should expect to see the red wire in RC/RH, the blue wire in C, the white wire in W, the yellow wire in Y, and the green wire in G. If the condenser unit is a heat pump, then there'll be a sixth wire. It'll be an orange wire in O/B. The blue wire in C is not always present because it's not required if the thermostat has batteries. If you find that the thermostat doesn't have any wire going to the C terminal, you should definitely put the blue wire there. Now go to the control board with the mindset of matching the wiring from the thermostat to the wiring at the control board. This means at the control board, you should have the red wire going to R, the blue wire going to C, the white wire going to W, the yellow wire going to Y, and the green wire going to G. Again, if the condenser unit is a heat pump, then you'll need to put the orange wire in O/B.

Now the black wires from the SS2 float switch and the two wires from the condenser unit work in conjunction with each other. Typically, the colors of the two condenser wires are red and white. This white wire goes into the C terminal on the control board right along with the blue wire that should already be there. The red wire of these two wires, coming from the condenser, has to be wire-nutted to either one of the SS2 float switch wires. The remaining float switch wire should be placed in the Y terminal right along with the yellow wire from the thermostat, which should already be there. At this point, all low-voltage wiring at the AHU is complete.

The last step to complete is mastic. Mastic is a vapor sealant, commonly referred to as "pookie," that is used to ensure that none of the cooling or heating vapors escape the ductwork. So, mastic has to be applied at any point ductwork connects to another piece of duct or the Ahu itself. Mastic is only used for sealing ductwork to another piece of duct or the AHU. Once you mastic all of the duct connections, you've completed the installation of the indoor unit.

Chapter 4
Equipment In: Outdoor Unit

N ow it's time to move on to the outdoor unit. The very first thing that has to be done is to set the pad that the condenser unit will be set on. You want this pad to be as level as possible so that the condenser unit can be as level as possible. To get the pad level, you need to use a shovel and a rake to even out the area of the ground where the pad will sit. You need to keep in mind that condenser units have clearance requirements that have to be adhered to. For example, Lennox brand condenser units require they be at least six inches from a wall. Daikin brand condenser units require ten inches. You should confirm what the clearance requirements are before setting the pad. Once you've set the pad in accordance with the clearance requirements, set the condenser unit on the pad. Set your torpedo level on top of the condenser and make sure it's level. If it's not level, you'll need to either manipulate the dirt underneath the pad or find something to put under the condenser to level it off. Once the unit is set, the first thing I like to do is fit the existing refrigerant lines into the unit. The same process you went through to fit the refrigerant lines into the AHU is the same process you'll employ to fit the refrigerant lines into the CU. There's no set way this has to be done. You really have to use your own discretion to complete this task. The thing you have to keep in mind when fitting the existing refrigerant lines back into the CU is that it should look uniform, meaning both the liquid and suction lines should be tight together and take up as little space as possible. You'll need to use things like ninety-degree elbows and couplers along with your copper cutters and copper benders to position the existing refrigerant lines to be fit into the CU.

After you complete that task, move on to the high-voltage wiring. During the disassembly phase, you disconnected a ground wire, a hot, and a neutral wire that were tied into the contactor. These three wires should be coming from the disconnect box outside, which we discussed in the disassembly chapter. It's time to wire those same three wires back up into the new CU. First, you'll need to remove the wiring panel door off of the CU using your 5/16" chuck on your impact drill. This will expose the contactor, the low-voltage wiring, and all of the other electrical components in the condenser. It's worth noting here that when I say, "wiring panel," I'm referring to the entire space that houses all of the electrical components of the CU. The panel door is only attached with two screws and can be taken off and put on with ease. The rest of the wiring panel should never be detached from the CU itself. Now, the first step in reattaching the high-voltage wires is to run the wires up through the allotted hole in the base of the wiring panel. Along the wall of the wiring panel, you'll see a small metal block with a green flathead screw in it. You can't miss it because it'll be the only green screw inside the entire panel. This is where the ground wire goes. Again, the ground wire is the only green wire of the three high-voltage wires. Use your multidriver to open up the green screw in the metal block. Slide the ground wire in, then use the multidriver to tighten the screw back onto the ground wire. Next, you'll need to put the other two high-voltage wires into the bottom of the contactor. It's literally the same exact process as when you disassembled the high voltage earlier. You'll need your Phillips head screwdriver now. Unscrew the Phillips head screw in the bottom of the contactor just enough to slide one of the wires in. If you look closely at the space where the screw is at the bottom of the contactor, you'll see there's a silver plate attached to the Phillips head screwdriver and then a gold plate that the screw is screwed into. Loosening the screw opens up the gap between the silver and gold plates. That gap is what you're sliding the wire into. It doesn't matter which of the two remaining wires you slide into that gap on the bottom of the contactor, so long as it's not the ground wire. Tighten the Phillips head screw back down onto the wire, and you're good to move on to the next. At this point, you should have one high-voltage wire remaining. This wire is going into the bottom of the contactor right next to the wire you just put in. That means you should see two Phillips head screws at the bottom of the contactor. Obviously, one of them is already holding one of the high-voltage wires. Put the last high-voltage wire behind the Phillips head screw right next to that one, and your high-voltage wiring is complete. At this point, you should have one high-voltage wire tied into the bottom left of the contactor, one high-voltage wire tied into the bottom right of the contactor, and the high-voltage ground wire tied into the small metal box with the green screw in it.

The last task to knock out in this particular step is wiring up the low voltage. Low-voltage wiring literally requires you to wire-nut two wires. Let's start by identifying the wires you're working with. So, do you remember the red and white wires that run alongside the refrigerant lines? If not, then you can identify it as the only wire coming out of the same exact wall penetration as the refrigerant lines. There is an allotted hole in the base of the wiring panel specifically for the low voltage wiring. You'll need to run your red and white wires up through the base of the wiring panel. Usually, there's a yellow and blue wire or either a black wire sitting right there next to the hole. If you don't see the wires sitting right there, then go back to the contactor. Look on the left and right sides of the contactor, and you'll see silver prongs on both sides. These prongs will have a yellow wire attached by a spade clip on the left side of the contactor and a blue or black wire attached to the prong on the right side of the contactor. Follow these wires to their end. Use a small blue wire nut to tie the yellow wire to the red wire. Use another small blue wire nut to tie the blue or black wire to the white wire.

The next task to be completed is the brazing process. Brazing is pretty much the same as welding, the difference being the temperature of the flame being used and the materials being bonded together. Brazing is the process of using gas, a torch, and a brazing rod to actually seal the refrigerant lines to the AHU and CU. As a beginner, you won't be the one doing any of the brazing, so we won't really cover it at this time. You should at least be familiar with the term, though. For now, you're all set as far as getting the CU installed. Now it's time to get the system ready to release the refrigerant.

Chapter 5

Releasing Refrigerant

Now that all of the equipment has been installed, it is time to work towards releasing the refrigerant. The first step towards releasing the refrigerant is doing what's called a pressure test. A pressure test is what tells you whether or not the refrigerant lines will be able to safely hold the refrigerant without any leaking out. After the brazing process is complete, the pressure test will tell you if every connection on the refrigerant lines is completely sealed. I mean, think about it, you had to connect the existing suction line to the suction line on the AHU and the CU. Then, that had to be brazed so that those suction line connections were completely sealed. If you used any ninety-degree elbows or couplers, then those connections should be sealed as well. So, you have to know for sure that all of these connections or joints will hold up under the pressure of refrigerant. Again, that's the whole purpose for doing a pressure test. The only things you need to complete a pressure test are a set of digital gauges, a tank of nitrogen, and a nitrogen regulator. I feel compelled at this time to emphasize a point. As a beginner you should own your own set of digital gauges with hoses (preferably one red, one yellow, and one blue).

As an employee, you'll never be expected to provide your own nitrogen because that's an item on the materials list for your employer. A nitrogen regulator is a completely different ball game, though. The nitrogen regulator is a tool/component that attaches to the nitrogen tank and regulates just how much nitrogen is allowed to leave the tank at a time. As a beginner, you won't be expected to have your own nitrogen regulator, but as you grow in the HVAC trade, you'll eventually realize that you're better off having your own rather than depending on someone else to have one. Some digital gauges have three ports, and some have four ports. Assuming your nitrogen regulator is already attached to the nitrogen tank, the first thing you'll want to do is hook one end of the yellow hose up to the nitrogen regulator. The other end of the yellow hose should still be attached to the middle part of your gauges if you have three ports. If your gauges have four ports, then the yellow hose should be attached to the third port from the left if the screen is facing you. Next, hook up your blue hose to the suction line port on the condenser unit. The other end of your blue hose should still be connected to the port on your gauges furthest to the left. It should even be labeled "LOW SIDE." Now hook the red hose up to the liquid line port. The other end of the red hose should still be attached to the port on your gauges furthest to the right. It should even be labeled "HIGH SIDE." Take a step back and make sure you have everything hooked up correctly. If your gauges have three ports, then your hoses should be blue, yellow, then red, going from left to right with the screen facing you. If it has four ports, then it should be blue, then an open port with no hose at all, then yellow, then red. From there, the blue hose should be attached to the suction line port, the red hose attached to the liquid line port, and the yellow hose attached to the nitrogen regulator. When you've double checked that everything is hooked up properly, it's time to pump nitrogen into the lineset. Start by opening up the knobs on the far left and far right of your gauges. These knobs are opening up the "LOW and HIGH SIDE" passageways on your gauges.

Opening these knobs is what allows nitrogen to pass through the blue and red hoses and into the suction and liquid lines. Next, open the nitrogen tank. This will allow nitrogen to go from the tank into the yellow hose. However, the port on your gauges that the yellow hose is attached to should still be closed at this time. As soon as you turn the knob to open this port, nitrogen will begin to rush through the blue and red hoses and into the lineset. The knob that opens and closes the port that the yellow hose is attached to is what you're going to use to control the flow of nitrogen. The nitrogen gas will flow into the lineset relatively quickly, so you should never walk away or leave an open tank of nitrogen unattended. Open up the part that the yellow hose is attached to and allow nitrogen to begin to fill the lineset. What you need to pay attention to are the numbers at the very top of your gauges. The numbers at the very top of your gauges represent the psig, which is the pounds per square inch, inside the lineset.

That's basically just a measurement of how much pressure is inside the lineset. You want to charge up the lineset to anywhere between 300-400 psig. Three hundred to four hundred psig is a good range because that's a far estimate of how high the refrigerant pressure will get. For the sake of simplicity, let's say you charge up the lineset to 350 psig with nitrogen. When your gauges read 350 psig close off the port the yellow hose is attached to. This should stop nitrogen from continuing to enter the lineset. Next, wait about five minutes and let the psig numbers stabilize. After about five minutes of stabilizing, the 350 psig will likely go to about 347 psig.

Whatever psig your gauges are reading after the first five minutes are the numbers you want to go with. Let's say after five minutes your gauges stabilize at 347 psig. You'll need to monitor the psig for twenty minutes. If that 347 psig drops more than two psig in twenty minutes, it's likely that the lineset has a leak. Losing more than one psig per ten minutes is an indication that there's a leak somewhere in the lineset. If you determine there's a leak in the lineset, just let the more experienced guy who's responsible for brazing know so he can correct the issue. Once the leak is corrected, or if the lineset sufficiently holds pressure, the next step is to vacuum the lineset.

As a beginner, you won't be expected to have your own vacuum pump but eventually you'll realize it's in your best interest to have your own. Vacuuming the lineset is the process of removing any moisture, air, and all other non-condensables out of the lineset. Vacuuming the lineset is super important because moisture and air are very harmful to the compressor. This may sound a little confusing because there is an actual liquid line and suction line present. For the sake of simplicity, for now we'll just acknowledge that the a/c system is designed in such a way that it doesn't allow liquid into the compressor. Getting liquid or moisture inside the compressor is the quickest way to destroy a compressor. So, the first step in avoiding that is to vacuum the lineset. To vacuum the lineset, all you need is a vacuum pump and digital gauges.

- Step one is to hook up the blue hose to the suction line port.
- Step two is to hook up the red hose to the liquid line port.
- Step three is to hook up the yellow hose to the vacuum pump.

If your digital gauges have four ports, the second port from the left will be labeled VAC. That port requires a special hose that hooks up to the vacuum pump. This hose is not required to pull a vacuum, but it will make the vacuum process go a bit faster. It's totally up to you whether or not you want to use this hose. Open up all of the ports on your gauges then turn the vacuum pump on. One point I should have emphasized at the beginning of the vacuuming process is it's extremely important to make sure the Schrader valves are back in place BEFORE you begin pulling a vacuum. Once you turn the vacuum on, the objective is to let it run until your gauges get below 500 microns. Micron is the unit that a vacuum is measured in. Keep in mind that vacuuming a lineset is the process of removing non condensables like moisture and air from the lineset. With that in mind, you can view microns as a measure of how many particles of moisture and air are present inside the lineset. You need to let the vacuum run until you have less than 500 particles of moisture and air left in the lineset. This process typically takes fifteen to thirty minutes in total but is contingent upon the quality of the vacuum pump, the pump oil, and the hoses. I recommend letting the vacuum run several minutes after it reaches 500 microns just so you get the microns as low as possible. Once you get the micron level down to a sufficient level, close off the HIGH and LOW side ports first (red and blue hoses at the gauges).

Then, close off the vacuum hose(s). Now you can turn off the vacuum pump altogether. Disconnect the vacuum pump because it's no longer needed at this point.

It's now time to release the refrigerant. The king valve caps should still be off. If they aren't, then remove them. Use your service wrench along with the service wrench hex adapter to release the refrigerant. The service wrench hex adapter attaches to the service wrench. Stick the other end of the hex adapter into the top of the king valve and turn it to the right. Do this for the liquid and suction lines. You'll literally hear the refrigerant rush into the lineset. You'll also be able to see the king valve rise to the top, the more you open it up. Next, put the king valve caps back on. You need to put new Armaflex insulation on the suction line now. The suction line always has to be insulated but the liquid line is never insulated. I always suggest putting new insulation on the suction line just for aesthetic purposes. If you have to reuse the old insulation, then do what you have to do.

Lastly, turn all the power back on. Make sure power is back on at the outdoor disconnect, the indoor breaker at the AHU, and the breaker inside the main breaker panel.

CONGRATULATIONS! You've completed a full a/c system installation. There's still a little more work to be done such as checking the refrigerant numbers, tweaking the performance of the system, and possibly adding refrigerant. These are tasks generally reserved for someone more experienced than a beginner. As a beginner, the only task remaining at this point is to clean up.

Chapter 6
Clean Up

The objective of the clean-up process is to make the work site look as though you were never there. That means removing all of the old equipment, disposing of all of the trash, and sweeping up all debris. Ensure all panels are re-attached to the equipment. You may not be required to, but you should also put silicone or caulk around the hole in the wall the refrigerant lines are penetrating through. Any type of drop cloths or floor savers should be picked up at this time. Make sure you leave the work site looking cleaner than you found it.

Conclusion

No, the HVAC industry is not for everyone, that's your advantage though. Just learning this information makes you hard to replace. Being successful in this field requires consistency, hard work, common sense, and a little physical strength. These are all traits that just about everyone coming out of a tough situation like being incarcerated already has. The characteristics you develop working for pennies while incarcerated, are the same characteristics the HVAC industry will pay you top dollar for. Lots of times, guys come home from incarceration, get whatever job they can, quit in less than a year because the effort they're putting in is not worth what they're getting paid, and end up back in the streets by year two. Learning and mastering the information in this book allows guys that are typically viewed as the black sheep to become the breadwinner in a relatively short period of time. And I'm totally speaking from experience on that.

Bio

I wouldn't consider my story quite a success story just yet but I am definitely on track to get there. I grew up as a little bit of a wild child in Jacksonville, Florida (Duuuuuuvvaaaaaaalll) with my mom, dad, and sister. I got myself into some trouble in 2006 and got sentenced to ten years in state prison about a week after I turned 21. I got out at the age of 31 and wasn't sure what my plan should be or what direction I should go in. I started off doing maintenance work for the Central Florida Fairgrounds. I knew that job wasn't going to get me the things I wanted in life so I began looking for other opportunities. After ten months of being back into society, I got a job as a helper for an HVAC company. They treated me like shit at first but I hung in there for one year and learned everything I could. One day I saw an invoice at work and I was able to see the entire breakdown of how the HVAC company I worked for was charging the client. My immediate response was

DAAAAAAAMMMMMN!!!! They're charging this company $38 an hour just for my services alone but they're only paying me $12. Right at that moment I started figuring out how I can become the person charging the $38 an hour rather than being the one getting paid the $12. I enrolled in a trade school for HVAC immediately. After being in school for about three months, my teacher got me a better paying job in HVAC. I went from making $12 to making $19 an hour in a little less than a year. As life continued to happen, I moved from Florida to Colorado. By that time, I had gained over two years of experience so I submitted my resume on INDEED and had a HVAC job waiting for me a month before I ever stepped foot in Colorado, paying $24 an hour. As of today, I'm still in Colorado making $28 an hour and I'm definitely underpaid (but we're working on fixing that). Now, I'm taking everything I've learned about the HVAC industry and teaching it to guys stepping back into society so they can immediately become an asset to their communities and families and earn themselves an above average living wage. I also offer an audiobook as well as an HVAC e-course for more in depth training. You can always access more of what I have to offer by texting HVAC to (201) 580-2192.

www.ingramcontent.com/pod-product-compliance
Lightning Source LLC
LaVergne TN
LVHW051748050326
832903LV00029B/2796